AF375202

Published by Enna's Press
ISBN: 979-8-9945835-4-8

Library of Congress Control Number: 2026903345

Art & Design: Created using licensed assets from Canva Pro.

For more books by the author, visit : iqralittlepages.com

JUVENILE NONFICTION / Science & Nature / Earth Sciences / Weather
JUVENILE NONFICTION / Science & Nature / Earth Sciences / Water

Printed in USA. First Edition 2026.

Note to Parents & Educators

Dear Grown-ups,

Little ones are born scientists, endlessly curious about the world around them. This book invites toddlers to wonder about a simple, daily miracle: "Where does rain come from?" We use the 4MAT learning cycle to guide young readers:

- *Why?* Connecting rain to their own joyful experience.
- *What?* Exploring the simple, concrete science.
- *How?* Modeling the cycle with a hands-on home experiment.
- *What else?* Seeing cycles everywhere in nature.

We blend this with Montessori principles, using realistic imagery, a focus on natural details, and a simple, respectful narrative to foster independent observation and discovery.

Discover more books for curious toddlers at : iqralittlepages.com

Happy exploring and wondering together!

Author & Parent

For Yahya

For the curious ones who look up at the sky
and wonder...

where does the rain come from?

You feel the rain on your skin. Drip! Drop!
It is cool and fresh.

You splash in the puddles it leaves behind.
Where does the rain come from?

Rain makes music. *Tap Tap Tap* on the window.
Plink Plink Plink on the leaves.
Splash in the puddle. Can you hear the rain's song?

After the rain, the air smells fresh and clean.
The earth is happy. The flowers are happy.
You are happy too.

The sun shines down, warm and strong

Its warmth lifts the water, light as air.

The tiny, invisible water droplets travel up and gather together. They form a cloud.

12

The cloud fills up... up... up... until it is heavy.
Then, the water falls back to the earth.
This is the rain.

13

The rain has a journey. It goes up to the clouds, and it falls down again. This is a cycle. It happens again and again.

We can see this journey in our home.
Steam rises, forms drops, and falls.
It is a small rain inside.

What if the water is in the big ocean?
Or on the sidewalk?
Or in your pet's bowl?
The sun's warmth still helps it on its journey up.

What if you watch the sky? The clouds gather.
The rain falls. The sun returns and sometimes... a
rainbow appears. Every rain has a beginning, a
middle, and an end.

Other things have cycles too. A seed grows into a plant, which makes new seeds. You have a cycle too: you grow, just like the plants and the rain.

19

Let's make a tiny rain storm inside!

Parents: This simple activity lets your child see the entire water cycle in action. Always provide close supervision.

Experiment Steps

1. **The Sun Warms the Ocean** - Hot water in jar
2. **The Cloud Gathers** - Plate on jar, steam collects
3. **The Rain Falls** - Ice on plate, droplets form and fall